An Introduction to Tree-Ring Dating

An Introduction to

TREE-RING DATING

Marvin A. Stokes and Terah L. Smiley

The University of Arizona Press

Tucson

The University of Arizona Press
© 1996 The Arizona Board of Regents
All rights reserved

www.uapress.arizona.edu

Library of Congress Cataloging-in-Publication Data
Stokes, Marvin A.
An introduction to tree-ring dating / Marvin A. Stokes and Terah
L. Smiley.
p.   cm.
Originally published: Chicago : University of Chicago Press,
1968.
Includes bibliographical references and index.
ISBN 0-8165-1680-4
1. Dendrochronology. I. Smiley, Terah L. (Terah Leroy), 1914– .
II. Title.
QK477.2.A6S88   1996
582.16'0372—dc20   96-30579

Manufactured in the United States of America on acid-free,
archival-quality paper containing a minimum of 30% post-
consumer waste and processed chlorine free.

16   15   14   13   12   11   8   7   6   5   4

Dedicated to

The People of the Navajo Tribe

CONTENTS

ILLUSTRATIONS

FOREWORD

In 1929 the University of Arizona's A. E. Douglass determined the exact ages of scores of prehistoric Indian ruins in the American Southwest using a novel dating technique based on the annual growth rings of trees. This event was widely heralded as a major scientific advance because it not only gave archaeologists their first real insight into the absolute chronological placement of the region's pre-Columbian cultures but also introduced an ingenious new method for establishing the precise calendar year date of an ancient wood or charcoal sample. Tree-ring dating is now routinely employed in many areas of the world and is still regarded as the most accurate dating method. Dendrochronology, as Douglass termed the overall field of tree-ring research, has become a flourishing scientific enterprise with a surprising diversity of direct applications in the social and natural sciences.

After the university established the Laboratory of Tree-Ring Research in 1937, one of its most productive applied research investigations was the Navajo Land Claim Project. This cooperative venture between the laboratory and the Navajo Tribe was initiated in 1951 under the guidance of Terah L. "Ted" Smiley, at that time the laboratory's curator of archaeological collections. Marvin A. Stokes joined Smiley two years later as a research assistant. The project entailed the collection and processing of several thousand tree-ring specimens from living trees, abandoned Navajo hogans, and other structures located both within and adjacent to the present Navajo Reservation. Well over a thousand hogan samples yielded tree-ring dates, which were then used as evidence by the tribe in its successful pursuit of compensatory claims presented before the United States Land Claims Commission.

I am quite certain that neither Smiley nor Stokes expected the Navajo Land Claim Project to go on intensively for eighteen years. I am equally sure that neither scientist anticipated that a significant by-product of the study would be the preparation of the book that was to become the classic introduction to the field of dendrochronology. Yet it is this unassuming account that has been the starting point for a generation of tree-ring enthusiasts, professional and amateur alike, and it is this slim volume that has consistently appeared in the tree-ring literature as one of the most frequently cited "how to" references.

Of course, the world of dendrochronology has changed dramatically in the three decades since this book was first published. There are now many more tree-ring research laboratories, a far greater number of practitioners, and a whole suite of new applications with fresh problems to be addressed. Nowhere are these changes more apparent than in the day-to-day operations in which tree-ring data are generated, processed, and analyzed. Yesterday's slide rule (see fig. 38) has been replaced by the math coprocessor of today's personal computer. But the reader should not be misled by these high-tech changes, because no matter how complex and sophisticated the tools and techniques of dendrochronology may become, the fundamental principles underlying the dating of a tree-ring specimen will always remain the same. Therein lies the beauty of this book, for the authors have presented the basic procedures of the entire dating process in such a clear and systematic fashion that their introduction is just as valuable to the interested reader today as it was nearly thirty years ago. To all who wish to understand better just how the tree-ring method works, I highly recommend this enduring book.

The senior author, Marvin Stokes, retired from the University of Arizona in 1989 and is now Professor Emeritus of Dendrochronology. Stokes conducted a number of tree-ring research projects in the Southwest and northern Mexico. Most notably, it was he who pioneered the use of dendrochronology in constructing reliable regional histories of past forest fires. Tree-ring-based fire history studies now constitute an important new field of investigation that has had a significant impact on the management of our nation's forest resources. For many years Stokes also taught the introductory course required of all dendrochronology students, and as those students will testify, his career was especially marked by the warmth and exceptional skills be brought to the classroom.

Ted Smiley played a leading role in the Laboratory of Tree-Ring Research in the early post–World War II period and served as its acting director for two years in the late 1950s. His primary interests, however, lay in the broader array of radioactive, geologic, climatic, and other dating methods, and it was he who organized and directed the university's innovative program in geochronology until it was merged with geology to form the present Department of Geosciences. Smiley was widely credited for his leadership in bringing national prominence to the Department of Geosciences by building it into one of this country's outstanding centers of Quaternary research. He retired from the University of Arizona in 1983 as a Professor Emeritus of Geosciences. Sadly, Ted Smiley died in February 1996, less than a year before the reissue of this book.

—Bryant Bannister

PREFACE

Dendrochronology, or tree-ring dating as it is often called, is defined as the study of the chronological sequence of annual growth rings in trees. Specifically, our concern in this book is with the task of establishing a calendar date for a wood or charcoal specimen.

.Tree-ring studies can be made in any part of the world where trees add an annual ring because of the genetic characteristics of the species. Most of the tree-ring work in the American Southwest has been on coniferous evergreens. Our attention in this area has centered on certain species of the pine family because they have wide geographic distribution, constitute a large population, and show excellent growth response to certain controlling factors. The photographs of trees and wood presented in this brief résumé are primarily of the common piñon pine (*Pinus edulis* Engelm).

This book is intended to give the reader an idea of what a dendrochronologist does in his studies of such trees. During the last several decades many people visiting the Laboratory of Tree-Ring Research have thought of dendrochronology as simply a counting of rings. Not a few of these visitors believed that when we sampled a living tree, we cut it down or injured it in some other way. Such is not the case, and we hope that this material, presented as a series of photographs supplemented by explanatory text, will clarify the process of tree-ring dating.

In this short introductory work, the basic principles of tree-ring dating are first explained, then details of the process are described, step by step, from the time the sample is collected in the field to the time it is finally dated and incorporated into a master chronology. Two points are stressed: the first concerns the necessity of systematically collecting samples and efficiently recording field data so that the results of laboratory analysis can be usefully applied; the second concerns the laboratory processes involved in dating specimens and compiling master chronologies. Although the emphasis is on dating "archaeological" specimens, and to a lesser extent on dendroclimatology, many of the techniques presented here are applicable to studies in other fields, such as botany, forestry, hydrology, and watershed management.

It is an axiom that in introductory studies explanations are often oversimplified for the sake of clarity. Here, for example, the relationship between ring growth and precipitation, and the anatomy and physiology of trees, are not so simple as the descriptions indicate. To any potential "do-it-yourself" dendrochronologists, our apologies for failing to discuss pitfalls and short cuts and for ignoring other details because of space limitations. Again, in laboratory analysis of data frequent use is made of statistical methods, and some of the basic steps (described in chapter 3) in constructing a master chronology can be accomplished by the use of a computer. These analytical methods, however, are beyond the scope of this book. We have included a bibliography at the end of the book to fill these gaps and to point the way to broader knowledge of the field of dendrochronology.

This book was made possible through the cooperation of personnel working for the Navajo Indian Tribe on their land claims case. This group was established to collect data for use as evidence by the Navajo Tribe in their case before the United States Indian Claims Commission. The Laboratory of Tree-Ring Research for a number of years was engaged in the dating of structural materials from old Navajo hogans as part of this work. This book was originally planned as an explanatory report for the people concerned in the claims case. We have included material relevant to other studies, but the nucleus of work in the land claims project remains.

The pictorial content of this volume is the backbone for the text. The largest contributor to this part remains Clifford

Gedekoh, field photographer for the Navajo Land Claim investigation. While his contribution, in the beginning, was much larger, what remains (Figs. 10, 11, 14, 15, 18, 20, 21, 22, 23, 24, 25, 28, 29, 36, and 38) is the nucleus for the presentation. To him go our thanks for his efforts. Our thanks also to Jeffrey S. Dean (Fig. 12) and Jack Hannah (Fig. 13) for the use of their photographs.

Special thanks must be made to Ruth Stokes and Marie Boyd for the endless hours of editing, typing, and critical review. Helen McQuay, Marianne Kelm, and Judy Bessler also helped at various stages. To our colleagues on the staff of the Laboratory of Tree-Ring Research, University of Arizona, our thanks for encouragement and suggestions.

M. A. S.
T. L. S.

Laboratory of Tree-Ring Research
University of Arizona
Tucson

An Introduction to Tree-Ring Dating

PRINCIPLES OF TREE-RING DATING

Introduction

The dendrochronologist is concerned with the study of the chronological sequence of tree rings. Dendrochronology is made possible by the fact that in many trees the annual rings visible in cross section, rather than all looking alike, exhibit characteristic patterns. Four conditions are necessary for these patterns to be usable in dating a specimen.

The first is that trees used for dating purposes must add only one ring for each growing season; hence we speak of the *annual* ring. Species which add more than one apparent ring during a growing season cannot at present be used for dating purposes.

The second condition is that although the total seasonal growth is the result of many interacting factors, such as genetics and environment, only one environmental factor must dominate in limiting the growth. In the American Southwest, this dominant limiting factor is precipitation. Elsewhere it may be something different. In Alaska, for example, it is temperature.

The third condition is that this growth-limiting climatic factor must vary in intensity from year to year and the resulting annual rings faithfully reflect such variation in their width. Although the ring width is not necessarily directly proportional to precipitation, the rings must be narrow in drought years and noticeably wider in rainy years.

It is this recognizable sequence of wide and narrow rings that makes possible cross dating, or the matching of ring patterns in one specimen with corresponding ring patterns in another. It has

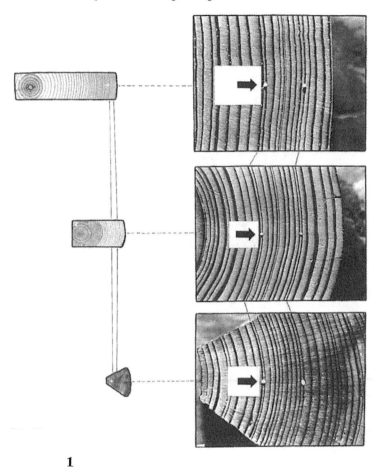

1

been observed that over a long enough period of time, the sequence of narrow and wide rings is never repeated exactly. For example, a dendrochronologist who is familiar with Southwestern chronologies readily recognizes the period A.D. 611 to 620 by its characteristic ring pattern and can match this pattern in one specimen with similar patterns from other specimens. Figure 1 shows this distinctive pattern in three different specimens, and

Figure 2 shows how the process of cross dating can be applied to specimens of different ages to produce a composite or master chronology covering a longer period of time than any of the individual specimens. In essence dating is accomplished by this pattern-matching, but as in many fields, the theory is simpler than the practice. Chapter 3 describes the mechanics of this process.

The fourth condition is that the variable environmental growth-limiting factor must be uniformly effective over a large geographical area. If this were not so, composite chronologies would have to be compiled for each small area. Minor differences, characteristic of small areas, always exist, but the basic ring patterns are similar enough to permit cross dating between trees growing many miles apart.

Cross Section of a Stem

A tree grows by increasing in height (apical growth) and by increasing in breadth (radial growth). This growth is the result of cell activity in meristem tissue in two regions of the plant. The apical meristem forms the primary tissue that causes the tree to extend the length of its stem and branches. The vascular cambium, derived from the lateral meristem, forms the secondary tissue that results in an increase in diameter. The vascular cambium divides in such a manner that cells formed to the inside of the cambium differentiate into the xylem, composing the woody part of the tree, and those formed to the outside of the cambium into the phloem.

The cross section in Figure 3 shows the xylem, marked by the annual rings, and the phloem, which appears as a dark band between the xylem and the bark. The cambium, which is located between the xylem and phloem, is not visible except with a microscope.

The phloem, which is essential to the tree because nutrients move through its cells, is of no use in dating, except indirectly, in that its presence on a specimen is assurance that no xylem is

6

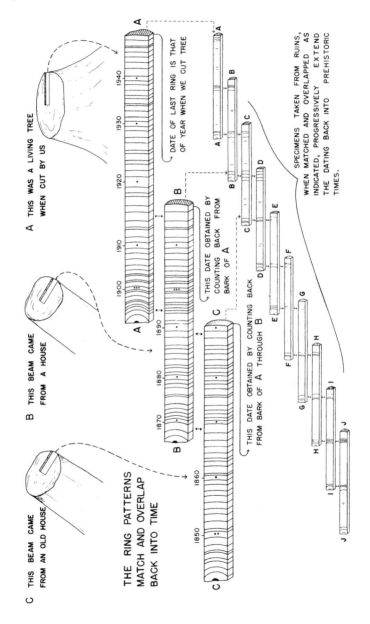

A THIS WAS A LIVING TREE WHEN CUT BY US

DATE OF LAST RING IS THAT OF YEAR WHEN WE CUT TREE

B THIS BEAM CAME FROM A HOUSE

THIS DATE OBTAINED BY COUNTING BACK FROM BARK OF A

C THIS BEAM CAME FROM AN OLD HOUSE

THE RING PATTERNS MATCH AND OVERLAP BACK INTO TIME

THIS DATE OBTAINED BY COUNTING BACK FROM BARK OF A THROUGH B

SPECIMENS TAKEN FROM RUINS, WHEN MATCHED AND OVERLAPPED AS INDICATED, PROGRESSIVELY EXTEND THE DATING BACK INTO PREHISTORIC TIMES.

1850 1860 1870 1880 1890 1900 1910 1920 1930 1940

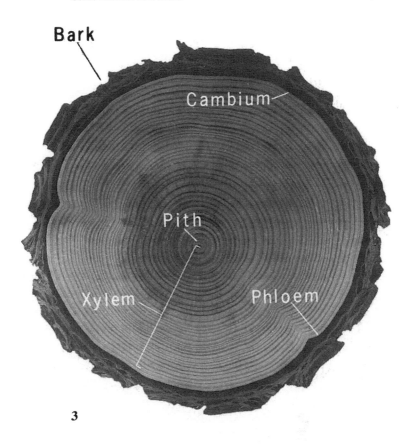

3

missing. The bark continually sloughs off so that only a few years
of growth remain, and this remaining portion has no ring structure
useful for dating.

After the primary tissue has been formed, the xylem growth
each year is laid down outside that of previous years and appears
as a ring in cross section as shown in Figure 3. Very crudely, one
might describe the annual growth layers as a series of progressively larger cones stacked one on top of the other. A cross section
removed at any level would show each cone as a ring and the pith
as a disk at the center.

Annual Ring Structure

The xylem in conifers is composed mostly of tracheid cells. A tracheid cell is long and thin and might be compared to a soda straw with both ends slightly tapered and closed. The long axes of the cells run parallel to the long axis of the stem or branch. In the microscopic cross section in Figure 4, the view is down into the cells, and they appear as openings almost rectangular in shape.

The annual ring is divided into two parts, earlywood and latewood. As the names imply, earlywood tracheids are formed at the beginning of each growing season and during the period of rapid radial growth, whereas latewood trachieds are formed toward the end of the growing season when cambial activity slows down. In latewood, tracheid walls are thick and strong and appear dark in color, and their cavities become progressively smaller. It is the sharp contrast between the last-formed latewood cells of one growing season and the first-formed earlywood cells of the following season that delineates the boundary of an annual ring. Because of the sharp contrast between the two cell types, annual

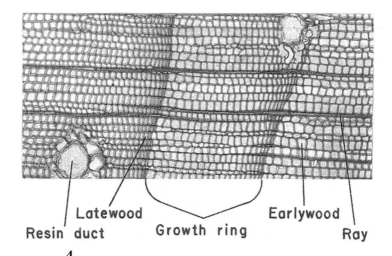

Latewood   Earlywood

Resin duct   Growth ring   Ray

4

rings can be seen in most cross sections without magnification.

The two large circular areas in Figure 4 are cross-sectional views of resin ducts. These ducts are found in resinous conifers, some species having only a few and others having many. Piñon pine (Fig. 4) has many resin ducts.

Radiating outward from the center of the stem are rows of cells whose long axes are at right angles to the tracheids. These cells, appearing as horizontal bands in Figure 4, are called rays. Their function is that of lateral conduction. They are of interest to the dendrochronologist primarily because they are an aid in identifying the species of wood.

The Effect of Site on Tree Growth

The growth of a tree is dependent on a complex series of inter-actions between genetic and environmental factors. The genetic makeup of the tree determines which environments the individual will tolerate and controls the response this tree will make to these environmental conditions. The environment supplies the nutrients, the water, and the radiant energy required for photo-synthetic and metabolic processes. The abundance, or lack, of any one or all of these constituents determines whether the tree will grow to the limits of its genetic potential.

It was stated earlier that precipitation is the dominant growth-limiting climatic factor in the Southwest and that growth varies with the amount of precipitation. This is essentially true if the trees to be sampled are chosen with care. More accurately, this growth-limiting factor is the effective soil moisture content, which is defined as the amount of available subsurface water coming from all sources minus that lost through evaporation and runoff. The amount of effective soil moisture is controlled not only by the amount, type, and timing of precipitation, but also by the texture, drainage, and composition of the soil.

If losses from runoff are low or if local underground water is available to the trees, the effective soil moisture content will be

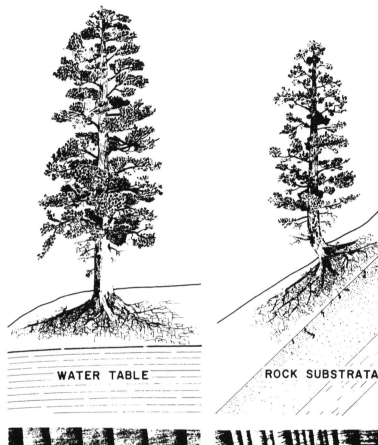

WATER TABLE

ROCK SUBSTRATA

COMPLACENT
RING SERIES

SENSITIVE
RING SERIES

sufficient in most years for a tree to produce optimum growth. When this occurs, the ring pattern is complacent—that is, there is insufficient variation in ring widths to produce any recognizable sequence. The sequence of rings may be uniformly wide or uniformly narrow. Figure 5 shows a typical complacent site and resultant ring series. Trees growing under these conditions may be excellent botanical specimens, but they are useless for dating purposes.

If sampling sites are selected so that no permanent underground water is available for growth and the soil drainage is good, radial growth is nearly enough proportional to total precipitation to produce datable ring patterns. Fortunately, the variation in total annual precipitation in the Southwest is great, which in turn results in appreciable variation in ring widths. Figure 5 also shows a sensitive ring series, obtained from a tree whose growth was controlled to a considerable degree by the variable condition of precipitation.

Through experience it has been found that generally lakeside, river valley, and roadside locations, as well as flat areas near slopes, produce trees with complacent rings. As a result, dendrochronologists are often found scurrying along rocky hillsides or clinging precariously to a tree with one hand to avoid falling down a steep slope, while removing a core sample.

Cross Dating

Atmospheric circulation, rainfall patterns, and mountain ranges divide the earth's surface into numerous "macroclimatic sites." Some areas, like the Sahara Desert in Africa, are large, and some, like the Olympic Peninsula rain forest in the northwestern United States, are small. In these climatic macrosites the annual meteorological conditions vary uniformly on a relative scale, and we consider each area to have, therefore, a homogeneous climate.

For example, in the Southwest a mountain may have a high annual precipitation on its crest and low annual precipitation at its

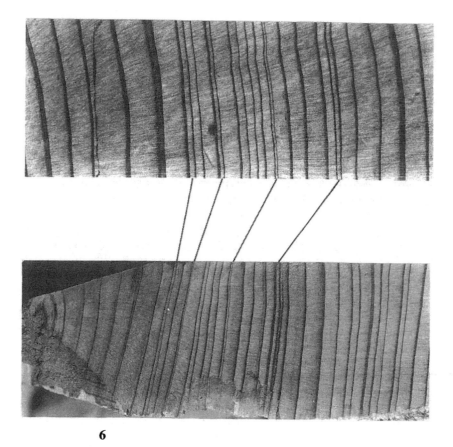

6

base. During a wet year, however, the precipitation will generally increase in both locations. As an added example, the annual precipitation in Tucson, Arizona, is approximately ten inches, whereas in Flagstaff, Arizona, 250 miles to the north, the precipitation is about twenty-five inches. If a "dry" year occurs, precipitation may drop to four or five inches in Tucson and to a proportional amount in Flagstaff.

Since the precipitation drop is roughly proportional throughout the climatic area, the ring patterns throughout this area should be similar. This has been found generally to be true with the result that trees within this area can be cross dated—that is, their ring patterns can be matched. Certain local or "microclimatic" differences are reflected in individual ring patterns, but usually these can be reconciled by adjustments.

It is perhaps necessary to point out that while all datable trees growing on sensitive sites within the climatic area produce similar patterns, the total growth differs greatly. As stated before, total precipitation varies greatly within the area; but other non-variable environmental characteristics, such as soil, also influence the amount of growth. Frequently, trees of unequal average ring widths are compared. The process of cross dating these specimens can perhaps be explained by the following hypothetical example.

Place a rubber band along the specimen having the narrower rings and mark on this band the ring-width pattern of this specimen. Place this band along the "longer" or wider-ringed specimen and stretch it until the band and the specimen are the same length. If the time periods are the same, the ring patterns on the band and the wood specimen can be seen to be essentially the same. There may be places where the match is not perfect, but supplemental stretching and shrinking of the band over these sections will provide a satisfactory correspondence.

Figure 6 shows two specimens which have been correctly cross dated.

Locally Absent Rings

One complication which sometimes arises in the process of cross dating is the absence of an annual ring at the location in the tree where the sample was taken.

A ring has been compared to a long, thin cone. The thickness of this "cone" is uniform neither in circumference nor along any line drawn on the stem; and therefore, the relative widths of rings at any place of sampling will vary slightly. Generally, the annual growth appreciably exceeds these variations so that the over-all ring pattern is not sufficiently different from sample to sample to complicate cross dating. The amount of total annual precipitation varies from year to year, and the growth response to this variation usually exceeds the differences between samples removed from the same tree.

Problems do arise, however, when rings of very dry years are encountered. A ring is formed every growing season (year), but in years of extremely little growth this ring may not show at every point on the cone. During such years, growth in the tree is likely to occur only at points of stress, such as the downhill side of a trunk or at a point near and under branches. Since these are the areas normally avoided in sampling, it is possible to obtain a core or cross section where a ring cannot be seen.

These "missing" rings can most easily be detected during the process of cross dating several specimens. The ring patterns will match ring-for-ring up to the year where a ring is missing in one of the samples. The ring count will be one year off after this point, unless correction is made by inserting a "ring" at the proper place in the sequence. Since a ring is never missing over the entire surface of a stem or branch, the term "locally absent" is used to denote a ring missing at the point sampled. The missing ring on the sample is marked by pricking the rings immediately preceding and following it (See Fig. 8).

Figure 7 (after Glock) diagrammatically illustrates the base portion of a tree stem. It shows three levels of cross-sectional surface, and each corresponding ring is connected with a vertical line. The ring representing 1847 is missing on the lowest section, appears as a lens between B and F and shows as a smaller ring in sections F and I.

Figure 8 is a schematic drawing of two specimens, which have been cross dated. The latewood is represented by lines and the earlywood by the spaces between the lines. Up to point A, cross-dating was done correctly by actually dating each ring by pattern recognition. The absent ring before point A, indicated by a broken line, was recognized, and the rings were carefully matched with this in mind. The solid lines drawn between the two plots

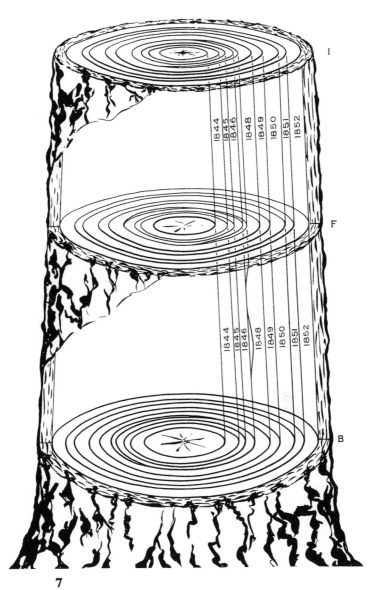

I

F

1844
1845
1846
1848
1849
1850
1851
1852

1844
1845
1846
1848
1849
1850
1851
1852

B

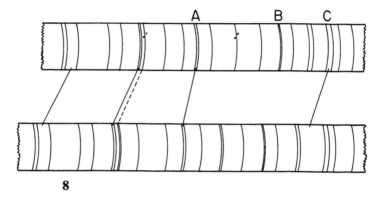

8

join rings with the same date. After point *A* rings were counted on each specimen and joined by a line (point *C*). Careful study of the pattern at point *C* shows that the rings no longer match and that the count is off (i.e., differs between the two specimens) by one year. This discrepancy occurred because the "absent" ring indicated on the first specimen between *A* and *B* was not compensated for in the ring count. Therefore, because of the occurrence of missing rings and other abnormalities, specimens cannot be dated by a simple ring count.

Double Rings

Another complication which arises in the process of cross dating is the occasional presence of "false" rings, or double rings, in the specimen being dated. The two terms are used interchangeably here because the effect is the same. A dark-colored latewood type of band appears in the light-colored earlywood of the ring (see Fig. 9). If this abnormality is not recognized in the dating process, the year's (season's) growth will be counted as two years and the ring count will be off by one year for each double ring overlooked.

There are several possible ways of detecting false rings. Fre-

DOUBLE

9

quently the last-formed latewood of a false ring is not clearly delineated because the latewood gradually blends in with the light-colored earlywood on either side. This gradual transition at the outer edge of a false ring, as contrasted with the abrupt change from latewood to earlywood in normal rings, is the most distinguishing characteristic of false rings and one which is fairly easily detected with a hand lens on a well prepared surface. If a cross section is available, a questionable ring can be traced around the entire circumference. If the latewood is discontinuous circumferentially, it is a false ring.

With somewhat higher magnification other diagnostic details become noticeable. Sometimes thin-walled earlywood type of cells can be seen to pass entirely through the false latewood. In specimens with resin ducts common in the latewood, one can observe that false latewood terminates at a duct while true latewood surrounds the duct incorporating it into the annual ring. These criteria for detecting false rings are useful for only a limited number of tree species, such as Douglas fir and a few pines, and are not always applicable for these. Some species (e.g., Arizona cypress and several types of junipers) have ring series in which it is often impossible to distinguish between true and false rings. When these methods of identification fail, false rings can sometimes be detected when specimens from the same site are cross dated ring-by-ring.

The causes of false rings are not well understood. Only a few species have been studied, and these not thoroughly. Abnormal climatic occurrences, such as a sudden, mid-growing season drought, have been suggested as possible causes; but this has been difficult to verify. Cross dating of several trees on a site has failed to show that double rings tend to be produced during certain years, as one would expect if the causes were climatic. On the other hand, climatic influences cannot be ruled out in the formation of false rings. Because false rings generally are not produced throughout the entire growth ring, it is possible to miss them in sampling either by core or cross section. This makes it difficult to correlate false ring production between trees and to correlate these rings with climatic variations. Currently, the role of auxin in production of false and true rings is being investigated. Some of the causes of false rings are apparently genetic because the tendency for production is more pronounced in some species than in others.

Termination of Growth

It is frequently desirable to be able to determine the date when the tree died from natural causes or from being cut. For example, this information is useful to an ecologist studying past environments or to an archaeologist attempting to determine the date that a structure was built. Unfortunately these cut or dead trees recovered in achaeological sites frequently have had their surfaces eroded so that an indeterminate amount of xylem is missing. In this situation the date of termination of growth cannot be determined, and all that can be said with certainty is that the tree died on or after the date of the outermost ring.

Sometimes there is evidence that the outermost ring on a specimen is the last one. The most conclusive evidence is, of course, the presence of bark. The outside date on the specimen in Figure 10 can positively be designated as the terminal or "cutting date" (if the tree was cut while alive and if the last-formed ring was not "absent"). To indicate this fact, a "B" is placed after the outside date on specimens retaining bark.

Good evidence that the outer ring is the terminal one or near to it is the presence of bark-beetle galleries, the "channels" in the specimen in Figure 10. These channels are made when beetles burrow into and through the soft, newly formed cells of the xylem and phloem. These insects attack recently killed trees that still have the bark or living trees that are weakened or dying from some cause. Since no healed galleries, which would be indicated by rings formed outside these galleries, have been found among the many thousands of trees studied in the Laboratory of Tree-Ring Research, we conclude that trees die soon after invasion by bark beetles. These galleries are usually only a few rings deep. Their presence indicates the outermost ring is very close to a true terminal date. When galleries are observed on a barkless specimen, this fact is noted with a "G" after the outside date.

The other two pieces of evidence for terminal growth are based on probability. If the outside ring extends around the entire circumference of a specimen, the probability is low that exactly the same number of rings eroded away around the entire specimen.

10

This evidence for terminal growth is noted with a "C" placed after the outside date.

Even in the absence of bark-beetle galleries, or of a continuous outer ring, it is sometimes possible to estimate terminal dates by comparing the outside dates of all of the wood samples from a single structure. Thus, if there are ten specimens from a single structure, all from different trees and all having nearly the same outside date, the probability is high that these dates are close to the terminal or cutting dates for these specimens. It is unlikely that all ten would erode or fracture so uniformly that exactly the same number of rings would be removed from each specimen.

2

FIELD METHODS

Collection of Archaeological Specimens

The Site

Material for tree-ring dating comes from many types of habitation sites. Examples from the American Southwest include a well-preserved hogan (Fig. 11), a pueblo with roofs intact (Fig. 12), and the charred and buried remains of a former dwelling (Fig. 13). The wood used in construction, and sometimes even the charcoal remains of a fireplace, offer a potential source of information on the occupation period of the site.

The field investigator examines the dwelling, noting the archaeological details. The floor and the nearby ground area are searched for potsherds, implements, and any other evidence of human occupation. As part of the regular procedure in archaeological research, specimens of wood and charcoal are collected and preserved.

Recording Site Information

A site data sheet is of the utmost importance. An archaeologist or student of history is interested in learning not only the time a site was used but also as much about the habits and life of the occupants as possible. A careful archaeological description (Fig. 14) is made of the horizontal and vertical stratigraphic position of the structure and of the relationship of the structure to the geographical area and to nearby structures. Some designation is assigned to the site, either a number or a name. This designation is then used as reference for all information on all specimens coming from that site.

11

12

13

Removing Samples from a Site

The same general procedures described here for selecting and removing samples from a Navajo hogan would apply equally well to any other type of structure or site for which dates are desired.

Several objectives must be kept in mind when selecting wood samples from a structure. Generally, only datable species are desirable—such species in the Southwest being, for example,

14

15

16

piñon pine, Douglas fir, and ponderosa pine. Samples of each kind of woody material found at the site should be preserved, however. Solid logs are selected, where possible, so that a complete ring pattern can be obtained; but any possibly datable specimen should be saved. Figure 15 shows men cutting a log from a hogan.

Some archaeological sites (e.g., see Fig. 13) contain no wood, or an insufficient amount for dating; but fortunately, some of these sites contain charcoal from burned structures or from camp fires. If this charcoal is of a datable species and the ring sequence is long enough, it can be dated in the same manner as wood. Charcoal or fragile wood, however, must be protected in some manner, usually by wrapping with cotton (Fig. 16). When it is possible in the field, charcoal is soaked in a paraffin-gasoline preservative. Recently, however, polyethylene glycol solutions have replaced this more volatile binder.

As soon as the specimen has been removed from the structure, the sample number, date, site number, location of the specimen

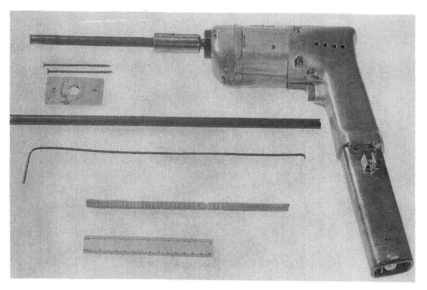

17

within the structure, and the space relationship of the specimen to other objects are recorded. This is necessary for a meaningful interpretation of its date.

Boring Tool for Archaeological Work

Sawing cross sections from beams is not always the most satisfactory method of obtaining archaeological specimens. Sawing is an arduous and time-consuming task, and sometimes can damage the rest of the structure (e.g., see the well preserved hogan in Fig. 11). To obviate this, special boring tools have been devised. Figure 17 shows one of these borers, which is driven by a battery-powered hand drill. The borer itself is a hollow tube with self-cleaning teeth set on the leading edge. Borer tubes of ¼ inch to 1 inch in diameter are used and come in various lengths, two of which are shown. The borer chuck is a special adaptation for this purpose. It is constructed so that the borer slips into the chuck and is held in position with a setscrew. A flat metal plate nailed to the timber holds the borer in position for the initial boring. When the desired depth of boring is reached, a long, hooked wire (see Fig. 17) is inserted between the core and borer tube and twisted slightly. This breaks the core from the rest of the wood and allows it to be removed. A core ½ inch in diameter with a sanded surface is shown in Figure 17.

Another alternative is a conventional electric drill powered by a 115-volt, a.c. portable generator. If a power source is not available, the end of this type of borer can be adapted to fit the chuck of a carpenter's brace. A. E. Douglass, originator of the science of dendrochronology, used such a tool in his early work in the Southwest.

Collection of Modern Specimens

Selecting the Tree

To date archaeological specimens, it is first necessary to have a master chronology (composite chronology) with which the specimens can be compared. At the beginning of the Navajo land claim study, a master chronology of the Colorado Plateau was already available and could have been used. It was considered desirable, however, to compile local, "regional" chronologies to check on any microclimatic conditions which could cause differences in growth within smaller areas. Also, regional chronologies must be kept up-to-date by continued collecting over the years. With current precipitation figures and other weather data available throughout the West, correlation of ring growth with measured climatic conditions has become increasingly possible.

Live trees used for compiling these regional chronologies were selected, when possible, from near the various archaeological sites to be dated (Fig. 18). Since the area under study comprised portions of the states of Arizona, Colorado, New Mexico, and Utah, this necessitated widespread collecting.

Trees from these sites were selected with several criteria in mind: Only datable species from sensitive sites were sampled.

18

Trees in dense stands were avoided because it has been found that competition among closely growing trees may modify or change the ring pattern from that of a "normal precipitation pattern." Healthy individuals with no obvious injury or disease, which could affect growth, were selected. Older trees in good condition are always highly desirable but not always available.

Increment Borer

The Swedish increment borer is a precision tool designed to remove a small core from a living tree without harming it. The small hole left in the tree after sampling is quickly sealed by sap, and only under extreme conditions can parasites or other forms of life enter the bore hole to cause damage.

The borer has a razor-sharp leading, or cutting, edge (Fig. 19, no. 3). The large screw threads behind this edge serve to draw the borer into the tree as the shaft is turned by the handle. The borer in Figure 19 removes a core approximately 15 inches long and $\frac{3}{16}$ of an inch in diameter. Borers up to 40 inches in length can be obtained. The 15-inch borer is generally used because it is convenient to carry in the field and is of sufficient length to reach the pith of most of the trees that we sample.

Coring the Tree

The tree selected is first examined to find the best place to remove a core. Areas where the ring patterns are likely to be distorted are avoided. These areas are most commonly found near branches or on the uphill and downhill sides of the trunk. When possible, samples are taken well below the first branches and on the side-slope segments of the trunk—that is, on the sides of the main stem that do *not* face uphill or downhill.

The tip of the assembled borer is pressed firmly against the bark at right angles to the axis of the trunk (Fig. 20), and the handles are turned clockwise. Once the borer tip is firmly anchored in the wood, pressure is needed only to turn the handles.

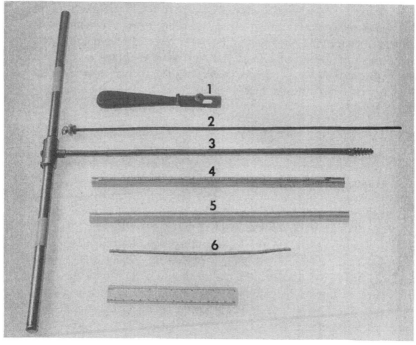

19

The borer is aimed at what the operator believes to be the pith, but this is not necessarily the center of the tree (i.e., trees growing on a steep slope will add more growth on the downhill side). Sometimes the operator will miss the pith and a few inside rings. These cores can be used, of course, but the time record will be incomplete. When the tree radius exceeds the borer length (e.g., in the giant sequoias), complete time records must be taken from cross sections of stumps left by logging. At the other extreme, if a tree is small enough, two cores may, in effect, be taken by boring through the center and out the other side.

Occasionally, a pitch pocket or decayed spot will be encountered while coring. This can be detected by observing the behavior

20

of the borer: when pitch pockets are encountered, the borer suddenly becomes much harder to turn; and when decayed spots are encountered, the borer spins very easily.

When either problem arises, the borer should be removed immediately. If boring into pitchy areas is continued, the borer can become so badly plugged that it is impossible to clean it in the field. The process of cleaning a badly plugged borer is too long to describe here. The best policy is prevention. A daily cleaning with kerosene is usually sufficient to remove pitch accumulated from coring normal trees. Pitch-free decayed areas do not cause these problems, but of course these cores are useless.

Care of the cutting tip of the borer should be taken, because it must be very sharp to work efficiently. The edge may be touched up occasionally with a whetstone, but any thorough sharpening must be done professionally.

Removing the Core

The borer is turned until the depth of penetration into the tree is sufficient to include the pith. The extractor spoon (Fig. 21) is then inserted into the borer from the handle end, so that it slides between the wood core and the metal sides of the borer. When the extractor is inserted to its full length, the borer is given a full turn counterclockwise to break the core from the tree; the extractor, carrying the core, is then removed from the borer (Fig. 21). The borer is withdrawn from the tree by turning the handle in a counterclockwise direction.

Handling the Core

Increment cores are fragile and thus must be handled with care. One method of handling cores (Fig. 22) is to fold a sheet of paper into an envelope. After the core has been removed from the spoon it is placed in this envelope, taking care not to break off the bark. The paper then is wrapped several times around the core to give added strength. Other storage methods involve the use of soda straws or corrugated cardboard. When the latter is

21

used, cores are numbered and laid between the corrugations, and the cardboard is then rolled into a bundle.

Information including place and date of sampling and field number is recorded on the container. This information is integrated into more complete field notes, which describe in detail the exact location of the site, direction or slope on which the

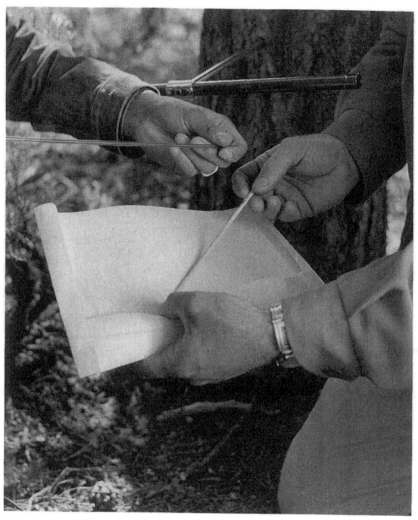

22

tree is growing, altitude, soil characteristics, associated species, relationship to other trees, and physical characteristics of the tree (such as stem diameter, height, and crown type).

LABORATORY TECHNIQUE

Preliminary Processing

When the samples are received in the laboratory, the number of each specimen is checked against catalogue information and against the duplicate field-data sheets, which accompany each group.

An index file card is completed for each specimen. On this card are recorded the sample number, identification of site or area from which it was taken, and the place of storage in the laboratory. The size, shape, species of wood, type of sample (e.g., modern core, archaeological wood, charcoal), condition of sample, and any evidence for the presence of the terminal ring are also recorded.

Dates and other data derived from studies are later added to the cards. The cards are stored in a master file which facilitates the finding of data on any of the laboratory specimens, which at present number over 80,000.

Preparation of Archaeological Specimens for Dating

Cutting the Bulk Wood Specimen: Figure 23

An obvious step in dendrochronological research is preparation of the material to best illustrate the anatomical characteristics desired for study. With large pieces of wood the specimen is examined to determine the best place for sampling, and a thin (about ½-inch) cross section is cut from that portion of the specimen.

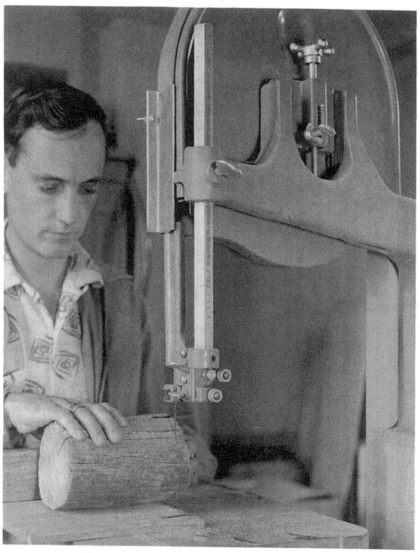

23

24

In selecting the area for cross sections, one must avoid cutting through a branch knot because of the distorted ring pattern in that area. Areas around the outside of the specimen with obvious wood loss also should be avoided. To be able to determine the *true* cutting date, it is best to cut either where bark is present or where beetle galleries are evident.

After the cross section is removed, the remaining portion of the specimen is placed in storage.

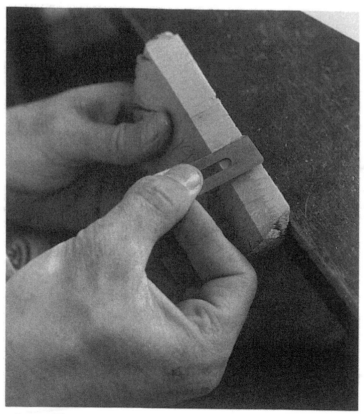

25

The Secondary Cut

After the cross section has been prepared, it is often desirable to make a secondary cut. A line intersecting the pith is drawn across the cross section at the maximum diameter. The cut is then made along this line at a forty-five degree angle (see Fig. 24). The secondary cut is made in this way because the tube-like tracheid cells (oriented vertically in the stem) are less likely to be torn or crushed when they are cut at an angle instead of in cross section. Also, subsequent slicing with a razor blade is much more satisfactory when done at an angle to the tracheids.

In spite of the advantages of the angle cut, it is sometimes preferable to use the entire cross section produced by the primary cut. Using an entire cross section increases the chance of detecting absent and double rings and other local distortions. With a cross section one can trace around the circumference at any ring to see if locally absent rings can be detected (see ch. 1, "Locally Absent Rings"). Rings suspected of being "doubles" can be followed around to see if ring characteristics (as described in ch. 1, "Double Rings") for doubles are visible. It also increases the certainty of finding the outermost ring on archaeological specimens. This is important because frequently several rings have eroded away.

Surfacing Wood Specimens

The surface produced by sawing a wood specimen is too rough to show adequate detail in ring structure; so further surfacing is necessary before dating can be done.

The quickest and most frequently used method of surfacing the secondary angle cuts is to slice the surface with a razor blade (Fig. 25).

When it is necessary to work with the entire cross section, the surface is best prepared by sanding with progressively finer grades of abrasive paper (No. 60 through No. 400) mounted on a mechanical hand sander. When the sanding is finished, cell walls should be readily visible and ring detail quite prominent (Fig. 26). All cross sections illustrated in this book were prepared in this manner.

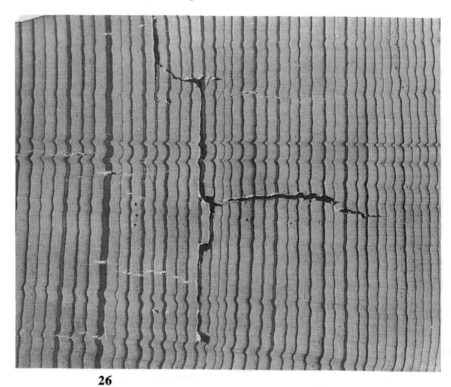

26

Preparing and Surfacing Charcoal Specimens

Since charcoal is fragile, and specimens are seldom large, the best procedure is to fracture the specimen to produce the desired surface and eliminate all the sawing and surfacing techniques described for wood. This is accomplished by taking the specimen in hand and breaking it at right angles to the long axis of the tracheids, just as one would snap a twig. When this technique is used, the break across the cell walls is clean and little or no debris remains in the cavities. The reflective surface of the charcoal is at its maximum efficiency (see Fig. 27), and ring detail is very prominent, even though the surfaced area is not a flat plane.

27

When it is absolutely necessary to surface a specimen, it must be treated with a paraffin-gasoline mixture (as described in Figs. 15 and 16), and a very sharp razor blade should be used. The cutting must be done with a gentle slicing motion or crushed particles of charcoal will fill the cell cavities and obliterate detail.

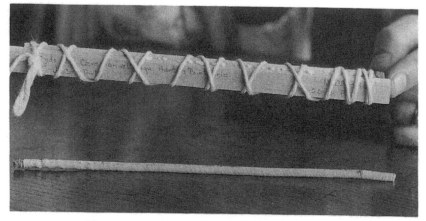

28

Preparation of Modern Specimens for Dating

Mounting the Core

Since a tree must be destroyed to obtain a cross section, almost all of our modern specimen are cores. Any cross sections we might use would be treated by the methods described for archaeological specimens.

Cores removed with an increment borer are small and fragile, and it is therefore necessary to mount them before any surfacing can be done. If the cores are still moist with tree fluids, they should be allowed to air dry for a few days before mounting, so that they will not shrink and pull apart in the mount as they dry. Drying cores in an oven is not advisable, since hastening the drying process increases the chances of breakage and extreme warping.

When the core has sufficiently dried, it is glued into a slotted mount, wrapped tightly with string (Fig. 28), and stored this way until the glue dries. The core should be oriented so that the observer looks down into the cells at an angle when the bark end is held to the right, following the convention of placing the most recent date to the right. This orientation is most easily done by

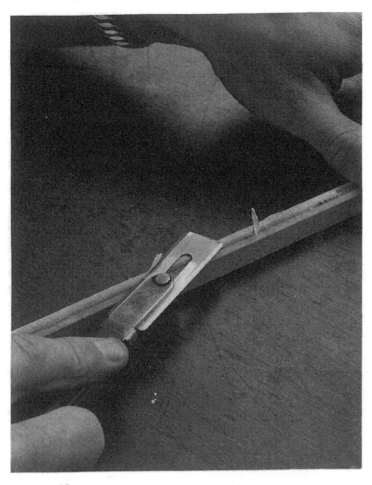

29

observing the innermost end of the core. If the pith is present, it will appear as a small, dark spot in cross section and as a vertical band in side view. If the pith is missing or the core extends beyond the pith, the orientation can be determined from the curvature of the innermost rings. These will appear as arcs in true cross section and as bands in side (to long axis of tracheids) view.

Surfacing the Core

After the glue is set and the string is removed, the core is surfaced for study—the quickest way being to slice the core with a razor blade as illustrated in Figure 29. The blade is held at an acute angle to the direction of slicing, and a flat surface is cut on the top of the core. If orientation of the core in the mount was properly done, this cut will be at an angle to the tracheid cross section.

Occasionally a core will contain a large amount of resin which makes it too tough to be properly surfaced with a blade. Often, soaking the mounted core in water will sufficiently soften the wood to allow a good cut to be made.

Excellent results can also be obtained by sanding the core in the manner described for cross sections. The first sanding, however, should be done with a finer grade of paper (No. 280 or No. 320). The flat surface is put on the core at this stage, and then it is sanded by hand with a very fine and clean grade of sandpaper mounted on a small sanding block. Figure 30 (top) shows a sanded specimen compared with an unsanded specimen. Note that both cores are in mounts.

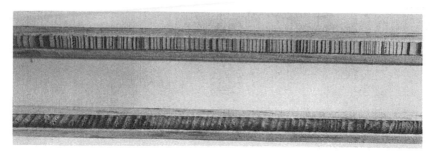

30

The Process of Dating Specimens

The Skeleton Plot

One of the major difficulties besetting any study is the reduction of all data to a form which can easily be used for analysis. Samples occupy a large amount of space, and are a nuisance to check each time information is needed. Specimens, even after proper surfacing, are difficult to compare directly one with another. Ideally, we reduce the information derived from study of these samples to paper in such a way that one specimen can easily be compared with another and so that data from several specimens can be combined to produce a suitable composite piece of information.

The skeleton-plot technique is one way in which data are reduced to paper. These plots are used as an aid for chronologically relating a group of specimens to each other by pattern matching and for determining dates for individual specimens of the group. The skeleton plot method has the advantage of being much faster than methods requiring actual ring measurement, but practice is required before the technique can be used effectively.

The process of dating is started by constructing a skeleton plot of each individual specimen. A strip of graph paper is labeled with the specimen number; and, to facilitate counting, a zero is placed at the extreme left of the paper, and every tenth square to the right is numbered. Each vertical line on the graph paper corresponds to one ring. The specimen is examined with a hand lens or other suitable low-power optical instrument so that all of the rings can be seen easily. The innermost ring on the specimen is plotted at zero, and the plotting progresses from this point outward on the specimen.

In skeleton plotting the narrow rings are the ones primarily being compared; so a line is marked at each interval where a narrow ring occurs. The decision of narrowness is based on the comparison of each ring with its immediate neighbors. The narrower the ring, the longer the line is drawn. The narrowest rings are arbitrarily represented with a line 2 cm in height, wide rings

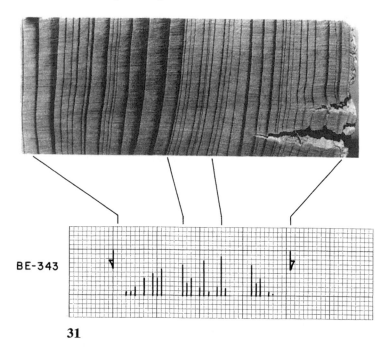

BE-343

31

are marked with a "B," and average rings are not marked. If a ring shows some slight reduction in width, but is not so narrow as to be indicated by a line on the skeleton plot, a dot is sometimes used to point out this fact.

Figure 31 shows a portion of archaeological specimen BE-343 with the corresponding portion of its skeleton plot. To facilitate comparison, a few rings on the specimen have been connected to their plots.

Skeleton plots of modern cores are prepared in the same manner as for archaeological specimens, except that the outside date of the core is known; and this date is marked at the right of the skeleton plot as soon as the plot is complete. It is necessary, however, to note whether the core was removed before, during, or after the growing season before assigning the outside date. For example, if a core was taken in March, 1961, before the growing season started, the outside ring would be dated 1960, not 1961. It is theoretically possible, of course, to core from a spot where the last-formed ring is double or absent, but an experienced dendrochronologist can soon detect this by checking the ring pattern.

The Composite Skeleton Plot

After each specimen in a group has been skeleton-plotted, several of these plots can be compared at one time. When this is done, similarities in their ring patterns can be noted and matched by placing the plots so these similar patterns are lined-up one under the other as illustrated in Figure 32. When this matching has been correctly done, all of the rings for any given year (although not yet assigned a date) will fall on the same vertical line. After all of the specimens have been lined-up, a piece of graph paper is placed at the bottom of the series, and a composite is made by plotting the average line length for each year. Since these lines are not measured, these averages, like the individual plots, are a matter of judgment.

The purpose of aligning the individual plots and constructing a composite plot is to find a time period common to all of the specimens. When this has been done, we say the specimens have been dated relative to each other but not yet placed in time. This process aids in the detection of abnormalities like missing or double rings, since the chance of an abnormality occurring in all of the specimens at the same year is small. The composite plot is usually longer than any of the single plots, which thus increases the chance of obtaining actual dates. When attempts to match skeleton plots prove to be unsuccessful, the original specimens are checked to detect any plotting errors. If none are found, it is assumed that the specimens have no patterns in common; and hence, they have no time period in common.

The theory of pattern matching has been discussed above (chap. 1, "Cross Dating"). Unfortunately, the actual practice is mastered by trial-and-error experience and cannot be adequately described. Careful examination of the three skeleton plots compared in

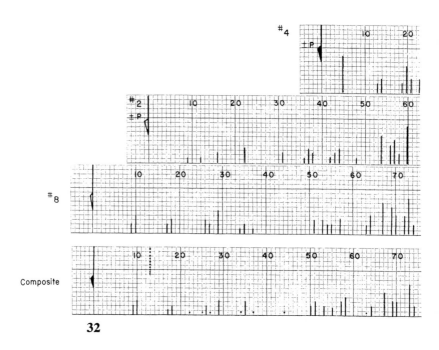

32

Figure 32 shows that, while several of the patterns match, there are many individual rings which do not match from plot-to-plot. This variation is typical. It is logical to ask how many such unmatched rings can be accepted in what we call matched plots. Our answer would have to be that, when *most* of the rings match, the fit is considered correct. While this may sound like a very unscientific answer, experienced dendrochronologists using these methods are able to duplicate each other. One helpful key to cross dating is to look for the time periods with very distinctive ring patterns; for example, 1890–1904. Experienced dendrochronologists carry many of these sequences in mind and can frequently date specimens from memory without resorting to skeleton plots.

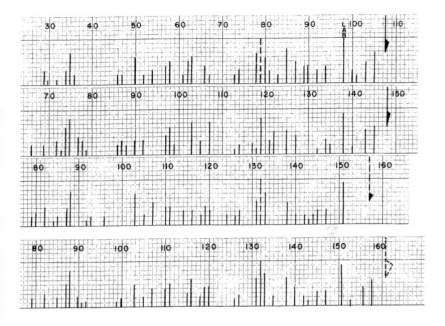

Dating the Specimens

When the composite skeleton plot for a group of archaeological specimens has been completed, the next step is to attempt to place the composite plot in time. This is accomplished by comparing the skeleton composite or a single skeleton plot with a plot of a master chronology (a dated composite chronology). The method of compiling a master chronology is discussed in detail later. The skeleton plot is moved along the master plot ring-by-ring until an alignment is found where the patterns match (see Fig 33). This technique is identical to that described in Figure 32 for matching skeleton plots of single specimens, except that the master plot is already dated. After it has been established that this is the only placement of the composite which produces a "match," the plots are compared on a ring-by-ring basis along the entire length of

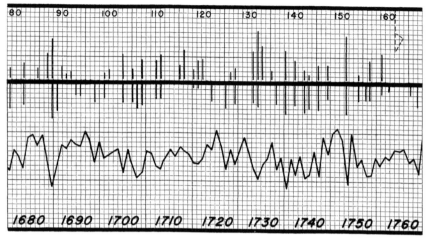

33

their overlapping portion. When this has been done, the composite has been tentatively placed in time. Each individual specimen must then be compared ring-by-ring with the master chronology before dates are considered to be verified.

In practice, very few of our cores are skeleton plotted before dating. If a series of cores is too difficult to date from memory, however, a few of the more promising ones are plotted and compared individually with the master plot; in this case the outside ring of the core has already been located in time.

Methods Used for Marking a Series of Dated Specimens

After the composite skeleton plot has been placed in time, each individual specimen in the group must be dated ring-by-ring. Usually, dating is started at a ring whose calendar equivalent is easily recognized and continues until every ring is dated. Dating every ring on the specimen, instead of just the outside one, not only serves to verify the outside date, but the information is used for other purposes, such as compiling master chronologies or for year-

34

by-year studies of climatic data. As the dating progresses, the rings on the specimen are marked as illustrated in Figure 34. Figure 35 gives the complete scheme of dating marks.

Compiling a Master Chronology

Basic Steps in Chronology Building

One of the ultimate aims of specimen analysis is to reduce the information obtained to some absolute form, which can be understood and used by others. While the skeleton-plot technique is an excellent tool for tentative dating, it is an unsatisfactory form for permanent storage or transmission of data. The construction of skeleton plots involves judgment, and the application of these plots is limited to the actual specimens plotted.

These limitations can be eliminated, if exact measurements are made of each ring width and if these measurements are plotted and ultimately converted to mathematical indices. Composite, or master, chronologies can be compiled by computing yearly averages of these indices. It is a simple matter to plot these yearly averages. By using this mathematical method, there is no limit to the number of specimens which can be "averaged" into a master

One pinprick indicates the DECADE.

Two pinpricks in a vertical alignment indicate the 50th YEAR.

Three pinpricks in a vertical alignment indicate the CENTURY YEAR.

Two pinpricks, horizontally aligned, indicate the presence of a "MICRO" RING.

Two pinpricks aligned at an angle across a latewood band indicate that a ring is MISSING from the sequence.

A SCHEMATIC RING SEQUENCE

1900   1910   1920   1940   1950

35

chronology, and the result is inherently more accurate than a composite skeleton plot. This process is described in detail in the sections which follow. Since the specimens used in constructing a master chronology have already been dated ring-by-ring, there is no further need to distinguish between archaeological and modern specimens.

Measuring Ring Widths of the Specimen

The specimen selected for measuring is mounted so that the ring structure is clearly visible in the microscope of the instrument being used. The magnification of the microscope in Figure 36 is ×40, which is adequate for a clear view of the individual cells in each ring. Cross hairs in the ocular serve as a reference point for measurement of ring widths. The microscope in Figure 36 is mounted on a stage, which can be moved along the stationary specimen by turning a crank so that the cross hairs in the ocular pass visually first over the earlywood of the ring and on to the outermost margin of the latewood of the same ring. The amount the microscope has moved (that is the ring width) registers on a dial. This value is read to the nearest .01 mm and is recorded on adding-machine tape.

There are other machines such as the Addo-X (Fig. 37), which automatically "reads" the distance the specimen has been moved relative to the microscope and records this value (to the nearest .01 mm) when a button is pushed. The process is faster, but the accuracy still depends on the operator.

All measurements are made along one continuous radius so that the relative ring widths are consistent. If it is necessary to shift the line of measurement when working on a specimen with an interrupted surface, each new radius is treated as an individual specimen.

Plotting Ring Widths and Growth Curve

After the specimen has been measured ring-by-ring, the next step is the plotting of these measurements. Standard sheets of metric-scale graph paper are glued together to form a sheet large enough

36

to accommodate all the desired plots. The specimens from a group are plotted on the same sheet with each vertical line representing one calendar year. The years are marked in decades, with the oldest date at the left side of the paper. The dates must span from the decade preceding the date of the innermost ring of the oldest

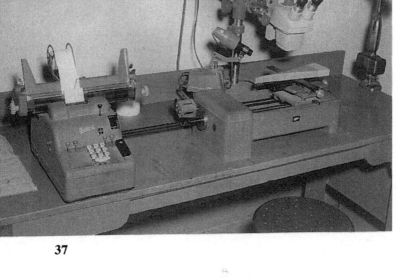

37

specimen to the decade following the outermost ring of the specimen with the most recent date. The usual vertical scale we use for plotting is 5 mm equal to 1 mm of measured ring width. If necessary, the scale for an unusual specimen can be reduced or enlarged so that all of the plots will be similar in magnitude.

The individual ring widths for each specimen are plotted as illustrated in Figure 38. To facilitate reading and later computations, the points marking the amount of growth for each year are joined by a line to form a "growth curve." These growth curves appear as jagged lines in Figure 38.

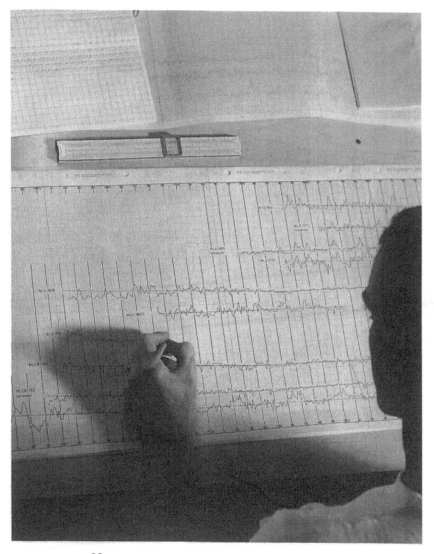

58

38

Constructing the Trend Line

It would be convenient if the master plot could be constructed by averaging all of the measured ring widths for each year and plotting these averages. This usually cannot be done, however, because of the growth characteristics of trees. A tree does not respond to a given set of environmental conditions in the same manner throughout its lifetime, and individual trees respond differently to the same set of conditions. Generally, a tree adds wider rings when it is young and the stem is small, and then the radial growth slows as the tree ages and the stem grows larger. A line describing this trend would have a sharply declining curve during the earlier years and a flatter curve for the later years. This growth tendency can be seen by studying the smooth lines in Figure 38.

Constructing a mean-growth curve or "trend line" for each plot is the next step in constructing a chronology. First it is necessary to compute a running-mean ring width. A 20-year mean is usually used for a long-ring series of 400 years or more, and a 10-year mean is used for a shorter series. To compute a 10-year running mean (the one used in Fig. 38), ring widths for each 10 years are averaged, and this average is plotted at the 5 and ½–year point of the interval. After all of the average points are plotted, they are used as a guide in drawing a trend line. Since this line is usually curved, however slightly, it is fitted by inspection. The trend lines appear as smooth curves in Figure 38.

Computing Annual Ring Indices

To be able to compare trees of different ages and growth rates, ring widths of a specimen are expressed in terms of the growth trend of that particular tree. In other words, one must be able to answer the question, "Is this ring wide or narrow for this particular tree at this time in its life?" This question is answered by comparing the plotted width of that ring with the "average" width (trend-line height) for that year. If the trend-line height is, for example, one unit and the ring width is 0.25 units, the index for that ring is 25 per cent or 0.25.

Not only is compiling a master chronology using indices more satisfactory than using actual values, but individual ring widths are more meaningful when expressed in percentage growth. It would be difficult, for example, to say, without studying an entire series at length, whether a ring 0.70 mm wide is large or small. There is no question, on the other hand, that a 20 per cent ring is narrow.

Constructing and Using the Master Chronology

The indices from a number of specimens from the same area may be averaged on a year-by-year basis. This series of average yearly indices constitutes a master chronology, and may be plotted (usually is) for use in dating additional specimens from that area, or nearby areas.

A plot of a master chronology is made on a long strip of graph paper using the usual horizontal scale of 1 square equals 1 year. The vertical axis is labeled from 0–200 (or whatever the high value was found to be in averaged indices) using a scale of 1 square equals 10 per cent. After all the points (yearly average indices) are plotted, they are joined together for easier reading. A skeleton plot can be made of the master chronology. In this case, however, more definite criteria are available; that is, if the ring falls below the 100 per cent line, a vertical line is inserted. Again, the narrower the ring, the longer the line will be. Figure 33 shows a portion of a master chronology with both skeleton- and connected-line points.

A "master chronology" may include specimens only from one small valley, or from one river drainage, or from one geographic unit. The essential feature is that annual-ring series making up the master chronology be homogeneous in character, the ring patterns resembling one another very closely. The similarity among annual-ring series, which constitutes the framework for building master chronologies, is sometimes easy to see but difficult to define. With the use of statistical analyses, correlation coefficients can be computed for ring series covering the same time period to provide a standard for expressing their degree of similarity.

Master chronologies are often made up by combining a number of ring series from smaller areas. Combining in such a fashion gives a fairly broad picture of tree growth over a wide area and is an aid in comparing ring records from regions within the total coverage of the master chronology. However, in some cases the smaller area chronologies are more useful to work with. For example, it is much easier to date specimens by the aid of a chronology made up of trees from the very same area than with one composed of trees which grew several hundred miles apart. The factors that make two ring series from adjacent areas similar and, therefore, cross datable, are the similar precipitation patterns. The farther apart the two areas are, the more dissimilar are the tree responses to the precipitation factors, and the more dissimilar are the ring patterns. Eventually, at a distance of about five to seven hundred miles, the patterns are so different that cross dating is not possible.

The body of text dealing with the techniques has already emphasized the importance of dendrochronology to the archaeologist. It is the only tool that offers a "precise" year-by-year chronology by means of which past events may be dated. In the Southwest, dated wood specimens from prehistoric ruins have been of great assistance to archaeologists attempting to interpret the cultural dynamics of those peoples without a written history.

In addition to its use in archaeology and climatology, dendrochronology promises to be a useful tool for the geologist studying time placement of erosion or alluviation processes.

Application of these techniques of tree-ring research to other species of plant growth is a field that remains largely unexplored. Some work has been done, again largely in the Southwest, but more remains to be done. Trees and shrubs that produce more than one definable growth layer per year should be investigated. Ideally, these investigations should be world wide rather than limited to the American Southwest.

Because tree-ring dating simply assigns a calendar year to the formation of a growth ring within a tree, however, the dating of any other event in relation to that tree ring will always remain an interpretive process and must be done in conjunction with additional evidence.

BIBLIOGRAPHY

A SELECTED BIBLIOGRAPHY

Bibliographies

Agerter, Sharlene R., and Glock, Waldo S. 1965. *An Annotated Bibliography of Tree Growth and Growth Rings. 1950–1962.* Tucson: The University of Arizona Press.

Kozlowski, Theodore T. 1958. *Tree Physiology Bibliography.* Washington, D. C.: Forest Service, U. S. Department of Agriculture.

Roughton, Robert D. 1962. "A review of literature on dendrochronology and age determination of woody plants," *State of Colorado Department of Game and Fish Technical Bulletin No. 15.* Fort Collins, Colorado: Colorado Cooperative Wildlife Research Unit, Colorado State University.

Archaeology

Douglass, A. E. 1935. "Dating Pueblo Bonito and other ruins of the Southwest," *Contributed Technical Papers, Pueblo Bonito Series No. 1.* Washington, D. C.: National Geographic Society.

Giddings, J. L. 1954. "Tree-Ring Dating in the American Arctic," *Tree-Ring Bulletin* 20 (3/4): 23–25.

Smiley, Terah L., Stubbs, Stanley A., and Bannister, Bryant. 1953. "A foundation for the dating of some late archaeological sites in the Rio Grande area, New Mexico: Based on studies in tree-ring methods and pottery analyses," *Laboratory of Tree-Ring Research Bulletin No. 6.* Tucson: University of Arizona.

Stokes, Marvin A., and Smiley, Terah L. 1963. "Dates from the Navajo land claim, I. Northern sector," *Tree-Ring Bulletin* 25 (3/4): 8–18.

Dendroclimatology and Cycles

Antevs, Ernst. 1953. "Tree-rings and seasons in past geologic eras," *Tree-Ring Bulletin* 20 (2): 17–19.

Douglass, A. E. 1919. *Climatic Cycles and Tree Growth*, vol. 1, no. 289, Washington, D. C.: Carnegie Institution of Washington.

———. 1928. *Climatic Cycles and Tree Growth*, vol. 2. Washington, D. C.: Carnegie Institution of Washington.

———. 1936. *Climatic Cycles and Tree Growth*, vol. 3. Washington, D. C.: Carnegie Institution of Washington.

———. 1936b. "Dendrochronology and studies in 'cyclics.' " *University of Pennsylvania Bicentennial Conference*, pp. 57–79.

———. 1937. "Tree-rings and chronology," *University of Arizona Bulletin* 8 (4): 1–36.

———. 1946. "Researches in dendrochronology," *Bulletin of the University of Utah* 37 (2): 1–19.

Fritts, H. C. 1965. "Tree-ring evidence for climatic changes in western North America," *Monthly Weather Review* 93 (7): 421–43.

Fritts, H. C., Smith, David G., and Stokes, M. A. 1965. "The biological model for paleoclimatic interpretation of Mesa Verde tree-ring series," *American Antiquity* 31 (2/2): 101–21.

Rudakov, V. E. 1958. "A method of studying the effect of climatic fluctuations on the width of annual rings in trees," *Botanicheskii Zhurnal* 43 (12): 1,708–12.

Schulman, Edmund. 1956. *Dendroclimatic Changes in Semiarid America.* Tucson: University of Arizona Press.

Smiley, Terah L. 1958. *The Geology and Dating of Sunset Crater, Flagstaff, Arizona. Black Mesa Basin.* R. Y. Anderson and J. W. Harshbarger, pp. 186–190. Albuquerque: New Mexico Geological Society.

General

Bannister, Bryant, and Smiley, Terah L. 1955. "Dendrochronology." In *Geochronology, Physical Science Bulletin No. 2*: ed. T. L. Smiley. Tucson: University of Arizona.

Glock, W. S. 1955. "Growth rings and climate," *Botanical Review* 21 (1/3): 73–188.

Schulman, Edmund. 1958. "Bristlecone pine, oldest known living thing," *National Geographic Magazine*, pp. 354–72.

Studhalter, R. A. 1955. "Some historical chapters," *Botanical Review* 21 (1/3): 1–72.

————. 1956. "Early history of cross-dating." *Tree-Ring Bulletin* 21 (1/4): 31–35.

Tree-Ring Bulletin. 1934–68. Complete set of volumes, 1–28, of the *Tree-Ring Bulletin* published by the Tree-Ring Society, Tucson, Arizona.

Techniques

Bannister, Bryant, 1962. "The interpretation of tree-ring dates," *American Antiquity* 27 (4): 508–14.

————. 1963. "Dendrochronology," pp. 161–76. In *Science in Archaeology*, ed. Don Brothwell and Eric Higgs. London: Thames and Hudson. New York: Basic Books.

Douglass, A. E. 1941. "Crossdating in dendrochronology," *Journal of Forestry* 39 (10): 825–31.

————. 1946. "Precision of ring dating in tree-ring chronologies," *University of Arizona Bulletin* 17 (3): 1–21.

Fritts, H. C. 1965. Dendrochronology, pp. 871–879. In *The Quaternary of the United States*, ed. H. E. Wright, Jr., and D. G. Frey. Princeton, New Jersey: Princeton University Press

Tree-Ring Studies

Bannister, Bryant and Scott, Stuart. 1964. "Dendrochronology in Mexico," *Actas y Memorias of Sobretiro del XXXV Congreso Internacional de Americanistas, Mexico 1962*, pp. 211–16.

Ferguson, C. W. 1964. *Annual Rings in Big Sagebrush, Artemisia tridentata*. Tucson: The University of Arizona Press.

Fritts, H. C., Smith, David G., Cardis, John W., and Budelsky, Carl. 1965. "Tree-ring characteristics along a vegetation gradient in northern Arizona," *Ecology* 46 (4): 393–401.

Fritts, H. C., Smith, David G., Budelsky, Carl A., and Cardis, John W. 1965. "The variability of ring characteristics within trees as shown by a reanalysis of four ponderosa pine," *Tree-Ring Bulletin* 27 (1–2): 3–18.

Giddings, J. L., Jr. 1941. "Dendrochronology in northern Alaska," *University of Alaska Publication*, vol. 5.

Glock, W. S., Germann, Paul J., and Agerter, Sharlene R. 1963. "Uniformity among growth layers in three ponderosa pine." *Smithsonian Institution Publication No. 4508*. Washington, D. C.: Smithsonian Institution.

Schulman, Edmund. 1941. "Precipitation records in California tree-rings," *Proceedings of the Sixth Pacific Science Congress, 1939*, pp. 707–17.

————1942. "Dendrochronology in pines of Arkansas," *Ecology* 23 (3): 309-18.

————. 1945. "Runoff histories in tree-rings of the Pacific Slope," *The Geographical Review* 35 (1): 59–73.

————. 1947. "Tree-ring hydrology in southern California," *University of Arizona Bulletin* 18 (3): 1–36.

————. 1951. "Tree-ring indices of rainfall, temperature, and river flow." *Compendium of Meteorology*, pp. 1,024–29. Boston: American Meteorological Society.

INDEX

INDEX